Published 2012 by Cliff Top Press Ltd.
www.arithmeticvillage.com

ISBN 978-0-9845731-8-9

Printed by Lightning Source, USA

Files licensed by www.depositphotos.com: Leather © montego • Background of burlap hessian sacking © odua
• Cardboard background © Leonardi • Fabric with red and pinks roses © mirabellart • Knit texture © Goodday
• Vintage pattern © Chamille White • Art vintage floral seamless pattern background © Irina_QQQ

Tina Times

By Kimberly Moore

To Minka

Terrific Tina
discovered the trick,

of adding in groups
to make counting quick.

Skipping through a garden, glistening a plenty,

five rows of four jewels, Tina has twenty.

She dines with five friends,
a ravenous bunch.

Thymes
Table
Menu

Two gems in each bowl,
means ten in their lunch!

She sees something twinkle, up in a pine.

Three groups of three gems that add up to nine.

When given flowers, she knows what to do,

seven flowers times six is forty two.

When six village cats slink across her floor,

a jewel on each paw
equals twenty four.

All of the answers, Tina knows by heart.

She's worked them all out on her times table chart!

People are different, each have a place.

Tina is loved for her speed and her grace.

Begin to multiply, try something new.

Count groups like Tina, it's easy to do!

Dear Grownup,

Tina Times personifies multiplication. As you read this book, children may spot the groups of jewels and begin to learn and understand the equations.

As numerical confidence increases, encourage your child to write down the equations for each page. Other children may wish to act like Tina and count in groups. You could even have a tea party to mimic Tina and her friends at lunch! Use this game to help fill out a times table chart.

With a light hearted approach, these experiences can support a lifetime love of multiplication!

For more inspirational ideas visit www.arithmeticvillage.com

A multiple of blessings,

Kimberly

Arithmetic Village

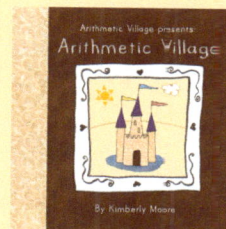

Polly Plus

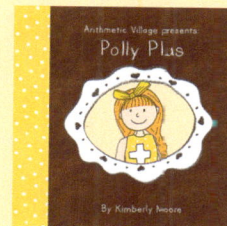

Linus Minus

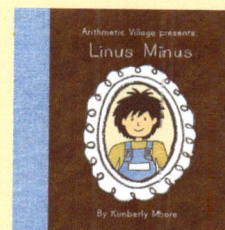

Tina Times

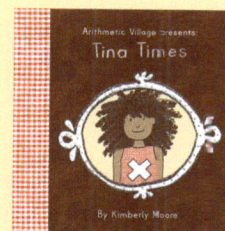

King David Divide

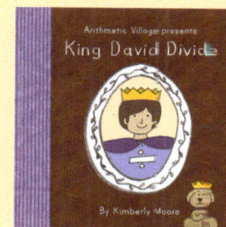